Beach Mysteries

ANNE HOTCHKIS

AuthorHouse™
1663 Liberty Drive
Bloomington, IN 47403
www.authorhouse.com
Phone: 1-800-839-8640

© 2013 Anne Hotchkis. All Rights Reserved.

No part of this book may be reproduced, stored in a retrieval system,
or transmitted by any means without the written permission of the author.

Published by AuthorHouse 04/30/2013

ISBN: 978-1-4817-0642-1 (sc)
978-1-4817-0643-8 (e)

Library of Congress Control Number: 2013900701

Any people depicted in stock imagery provided by Thinkstock are models,
and such images are being used for illustrative purposes only.
Certain stock imagery © Thinkstock.

This book is printed on acid-free paper.

Because of the dynamic nature of the Internet, any web addresses or links contained in this book may have changed
since publication and may no longer be valid. The views expressed in this work are solely those of the author and do not
necessarily reflect the views of the publisher, and the publisher hereby disclaims any responsibility for them.

This book is dedicated to
Jason, Josh and Meredith

Five Points

central disk
five arms radiate from it
I am a starfish of this beach

difficult to find
starfish can be found
under seaweed or
on rocky outcrops

they have suction cups on their underneath
which stick to rough stones

pick me up
hold me upside down
my suckers quiver in your hand
my five arms go straight and flat

flip me over and return me to my original home

Five Points

1. What kind of sea creature is this poem about?
   _____

2. Where on the beach can you find them?
   _____

3. When is the best time to look for them?
   _____

4. What are the suction cups for?
   _____

5. In your family who likes starfish?
   _____

6. What colour are starfish in the Northumberland Strait?
   _____
   _____

7. Why do they have five arms? *research
   _____
   _____

8. How would you describe a starfish?
   _____
   _____

9. Are they an endangered species?
   _____

10. Draw and colour five different-sized starfish of the Northumberland Strait.

Don't Pinch Me

don't pinch me
I can't see
the hidden crab
I fear for my Lab
his nose gets too close
the crab's claws

ATTACK

My Lab gives a YELP
for my help

stay away
go home
you don't belong here
while I beachcomb the tidepool

he leaves
with tail between
his legs

I better go
cheer him up
leave the beach
and the crab
for another day

Don't Pinch Me

1. What marine creature is this poem about?
   _____

2. What happened to the Lab pup?
   _____

3. Why did the crab attack the Lab?
   _____

4. What sound did the pup make?
   _____

5. Where does the Lab pup go after he has been pinched by the crab? _____
   _____

6. Do you have a pet dog?

   a) If yes, how would it react to a crab's pinch?
   _____

   b) If not, imagine a pet dog of your own and answer question
   _____

7. Draw a picture of a crab. *Research and label all claws and body parts.

Imprisonment to Freedom

hermit crab
laden down
hard house rests on back
weighted with pain like those imprisoned
search for peace
freedom.

Oh, hermit crab
so plentiful at ebb tide
we, too, are abundant
on this planet
seek a safe haven.

molted mollusk shell
stripped of home
hunt for roomier residence
fear of predators
identical plight.

individuals pray to higher self
meditate
deepen into faith
share oneness with others
give a gift of love.

hermit crab fulfills exploration
new habitat found
trepidation over
dilemma ends.

Imprisonment to Freedom

1. What sea creature is this poem about?
   _____
   _____

2. What do they look like?
   _____
   _____

3. Where do you find them?
   _____

4. Have you ever seen one?
   _____

5. What is the home of a hermit crab?
   _____

6. When does a hermit crab find a new home?
   _____
   _____

7. How does he do this?
   _____
   _____

8. Draw and colour a hermit crab poking out of its mollusk shell.

9. Define

   1.) habitat- _____
   2.) meditate- _____
   3.) predators- _____
   4.) trepidation- _____
   5.) dilemma- _____
   6.) mollusk- _____
   7.) molted- _____
   8.) abundant- _____
   9.) laden- _____
   10.) safe haven- _____

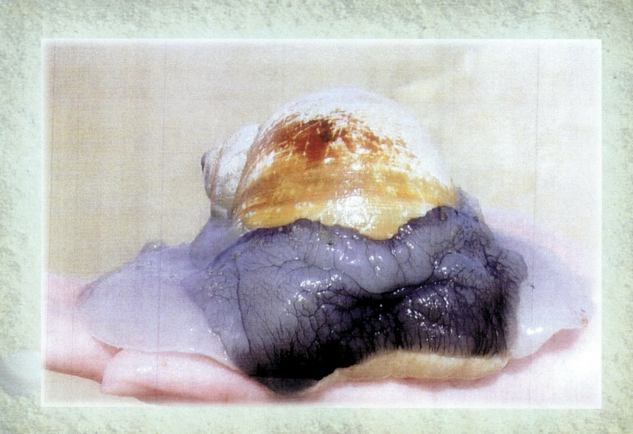

Sea Creeper

slow-moving mollusk
spirally-coiled shell
with a ventral muscular foot
covered in sand
seized in the ebb tide
on sand bars or
in gullies

litter on the intertidal zone
I collect exquisite colorful ones
to add to my beach shell treasures

I traverse the scenic, expansive red sand bars
search for live sea creatures
feel joy and wonder of it all

I traipse through the tidal pools
find many of these slow-moving shells
with their foot exposed to feed

I spy two of them—moon snails—on top of a crab
they suck the life out of it.

Sea Creeper

1. What is a moon snail?
   _____

2. What does it look like?
   _____
   _____

3. Where can you find them?
   _____

4. Describe the moon snail.
   _____
   _____
   _____

5. What is an intertidal zone?
   _____
   _____

6. How do moon snails move?
   _____
   _____

7. What is a ventral muscular foot?
   _____
   _____

8. When can you find them?

   _____

9. Define
   1.) seized- _____
   2.) exquisite- _____
   3.) traverse- _____
   4.) expansive- _____
   5.) traipse- _____

10. Draw and colour a picture of two moon snails feeding from the top of a crab.

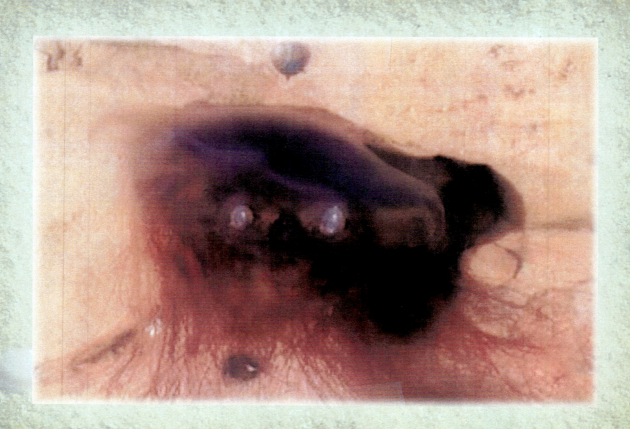

Purple Jello
Red Shoestring Licorice

wobblier wobbly woo
I seek jello patches with you
they stain the ebb tide sand
we cut
chop
and slice
jellyfish into pieces
the mutilated purple jello becomes purple soup
the red shoestring licorice dries up in the afternoon sun

tide flows inward
the damaged, severed jellyfish
multiply
into many newly-formed smaller jellos
red shoestring licorice stingers attach
to each and every one

Purple Jello
Red Shoestring Licorice

1. In your own words describe a jellyfish.
   _____
   _____
   _____
   _____

2. How does the poet describe a jellyfish?
   _____
   _____

3. In the poem, what happens to the jellyfish when the tide goes out? _____
   _____
   _____

4. What does the poet do to the jellyfish on the sandbars?
   _____
   _____

5. When the tide is in, what do you have to be careful of when you're swimming? _____
   _____

6. Define
    1.) mutilated- _____
    2.) stingers- _____
    3.) severed- _____
    4.) licorice- _____

7. Draw and colour two pictures.
    1.) jellyfish on a sandbar and
    2.) jellyfish in the water.

Ticklefish

tickle my toes
tickle my feet
tickle my hands
as I catch them buried in sand

they jump up
and down
from my legs
to my toes

SAND SHRIMPS

find safety in tidepools
at the ebb tide
put your feet in the gully
now

LAUGH

out loud

Ticklefish are a fun-filled marine sand shrimp with which to play

Ticklefish

1. What is the real name for ticklefish?
   _____

2. Have they ever tickled you?
   _____

3. Where can you find them?
   _____

4. What does ebb tide mean?
   _____

5. What does a ticklefish do to protect itself from predators?
   _____
   _____

6. Why does the poet say they are playful?
   _____

7. Define
   1.) predator- _____
   2.) camouflage- _____

3.) sand shrimps- _____

4.) marine- _____

5.) gully- _____

8. Draw and colour

   1.) A sand shrimp

   2.) A person being tickled by sand shrimps (ticklefish)

Underwater Magic

Look UP
Look DOWN
the undersea garden is bountiful
abundant with the COLOURS
TEXTURES of numerous marine plants
seaweeds wave in the ocean current
kelp
irish moss
sea lettuce
edible red lettuce
seaweed with bladders

scuba diving allows one to visit the ocean in 3D
above I see kelp float with the sparkle of sunshine
below various seaweeds sway in a gentle flow

I return to the beach as the tide recedes
seaweed lies flat upon shoreline rocks

Underwater Magic

1. Name five kinds of seaweed in this poem.
    1.) _____
    2.) _____
    3.) _____
    4.) _____
    5.) _____

2. How can one visit the ocean in 3D?
    _____

3. What happens to seaweed when the tide is out?
    _____

4. *Research
    a) In what foods do you find seaweed?
    _____
    _____

    b) What other uses are there for seaweed?
    _____
    _____

c) Write a paragraph in a scribbler about each of the seaweeds in the poem.

5. Draw a colourful picture using as many seaweeds as possible.

Fishing Without A Rod

trapped by the ebb of the tide in
gullies between sand bars
chubs like fresh water minnows
swim the boundaries of the tidal pool
smooth sand-colored fish
I leap into the school of chubs

they S C A T T E R

hide behind rocks
and under floating seaweed
quickly I make a V-shape with my feet
two chubs find refuge under my arches
I reach down and cup my hands around
them
my pail beside me

I have caught my first fish of the day

I proceed to follow the school
jump into the middle

V-shaped feet

capture two more fish in my arches
over and over and over until the catch is finished
the tide flows in
what happens to the pail of chubs
I take them to the wharf and use them as bait for bigger fish

Fishing Without a Rod

1. How does the poet fish without a fishing rod?
   _____
   _____

2. What is the name of the kind of fish in the poem?
   _____

3. Where do you find them? _____

4. When can you find them? _____

5. Describe what these fish look like.
   _____
   _____
   _____
   _____

6. Describe how to catch them.
   _____
   _____
   _____
   _____

7. What does making a V-shape with your feet do in the process of catching fish?
   _____

8. Why do the fish scatter?
   _____

9. Where do they go when they scatter?
   _____

10. What does the poet do with the catch of the day?
    _____

11. Draw and colour a school of chubs.

Delectable Delicacy

tide's out
beach comb the scraggly rocks at the points
collect a bucket full of mussels
the dark blue ones
clean them overnight
with cornmeal
cook tomorrow
steam in sea water
serve with melted butter

if you prefer, buy cultivated ones at the fish pound
great on the barbecue
best beach bivalve
scrumdiddlyumptious

Delectable Delicacy

1. This delectable delicacy is about _____.

2. What does the poet mean when she says "clean them overnight with cornmeal"? Explain
   _____
   _____

3. How are mussels cooked?
   _____
   _____

4. How do you serve them?
   _____
   _____

5. Why are mussels considered a delicacy?
   _____
   _____

6. Have you ever eaten mussels? _____

7. Have you ever cooked them on the BBQ?
   _____
   _____

8. What is a fish pound?
   _____

9. What does the word scrumdiddlyumptious mean?
   _____
   How many letters in this word? _____
   Can you memorize the spelling of this word? _____
   Can you use it in a sentence?
   _____

10. Define
    1.) scraggly- _____
    2.) cultivated- _____
    3.) bivalve- _____

11. Draw the scraggly rocks and add clusters of colourful mussels to them. Use a sheet of paper.

12. *Research. Find out how cultivated mussels grow. Write a paragraph about them in a scribbler.

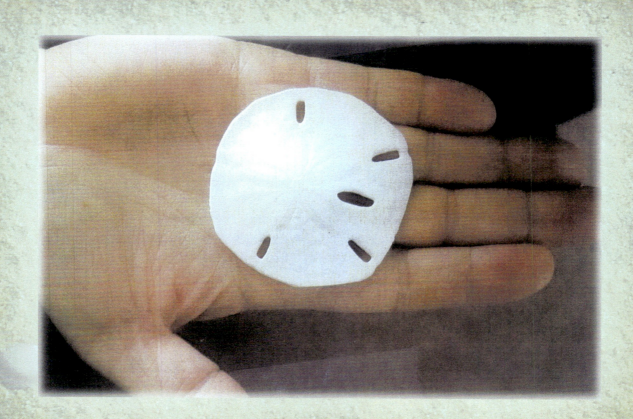

Money on the Beach—Oh yeah!

full moon
lowest tide
wade at water's edge
slow
careful
wander
search for sand dollars

Money on the Beach—Oh yeah!

1. What kind of marine animal is this poem about?
   _____

2. Where on the beach do you find them?
   _____

3. Have you ever found one?
   _____

4. Are they an endangered species on the Northumberland Strait?
   _____

5. 5. Why is this poem called, "Money on the Beach"?
   _____

6. Do you think this is a good title? Why? or Why not?
   _____

7. *Research—google-- Northumberland Strait sand dollars. Draw and colour as many kinds of sand dollars as is possible for this area. Use a sheet of paper.

8. When is the best time to find sand dollars?
   _____

9. *Research. What does the full moon have to do with finding sand dollars?
   _____

Open Door/Closed Door

crustaceans made of separate plates

fastened on rocks and ship bottoms

white, hard stuck like glue

flow of tide rises and advances

the little trap doors

open and expose

feathery fronds which sway
back and forth

back and forth

back and forth

to catch floating edible tidbits

the ebb of the tide

closes the barnacles

beachcombers tread delicately

on the crustaceans' homes.

Open Door/Closed Door

1. What does the title of this poem mean?
   _____
   _____

2. What is the name of this marine animal?
   _____

3. What colour are they? _____

4. Where do you find them? _____

5. What are the little trap doors?
   _____
   _____

6. How does a barnacle feed?
   _____
   _____

7. *Research. How do barnacles grow?
   Use a scribbler.

8. What does the tide have to do with barnacles?
   _____
   _____

9. Have you ever walked on barnacles?
   _____

10. Define
    1.) crustaceans- _____
    2.) edible- _____
    3.) feathery- _____
    4.) fronds- _____
    5.) tidbits- _____
    6.) tread- _____
    7.) delicately- _____

11. On a sheet of paper draw barnacles of different sizes growing on rocks.

12. On a sheet of paper draw a barnacle under water with its door open and its feathery fronds collecting edible tidbits.

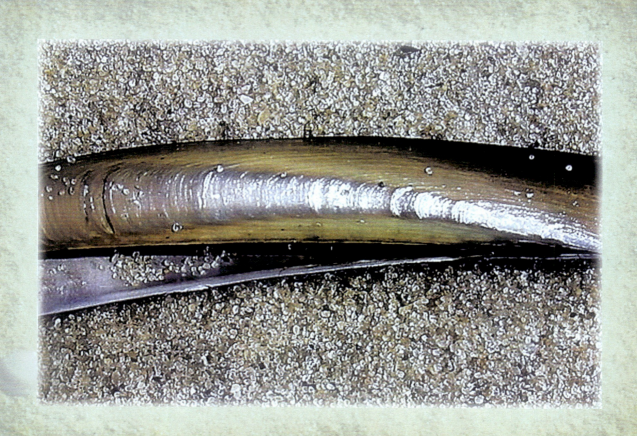

You Can't Shave With Me

elongated bivalve shells
razor sharp edges
an extended foot that burrows into the sand
leaving holes on the surface of the sandbar

razor clams are plentiful
but difficult to catch
find a gap and as quick as a wick
stick your finger in it
with luck you may touch the top end of the razor clam
push the edge of the shell against the side of the hole
dig it out

QUICK

Try try try over and over again
until you master this technique
great bait for wharf fishing

You Can't Shave With Me

1. What is a razor clam?
   _____

2. Is it easy to catch live ones?
   _____

3. Are they plentiful along the Northumberland Strait?
   _____

4. How do you capture one?
   _____
   _____

5. Define
   1.) elongated- _____
   2.) extended- _____
   3.) burrows- _____
   4.) technique- _____
   5.) bait- _____
   6.) wharf- _____

6. Draw and colour

   1.) a razor clam with its foot sticking out

   2.) an empty razor clam shell.

Tide's Out

quiver and quake
beneath the sandbars
crack like dandelions in paved driveways
appear on the furthermost tide flat
quahogs show their presence
pail and shovel
I begin my quest for a delectable treat
one, two, three…
I dig up the visible fissures in the sand
Four, five, six…
the search becomes a delightful adventure
filter feeders siphon water through their bivalve system
reach in my pocket
I tug the salt bag hidden there
I sprinkle a pinch of salt in between the two shells
the mollusk outstretches its foot to lick the invasive extra salt
my experiment proves a success
bucket full I return to the cottage in preparation for the nighttime
ritual
clear ocean tidal water collected
rinsed quahogs returned to pail
unmeasured cornmeal served to clams
siphon through the filter feed systems
discarded sand settles in the bottom of the pail
ready for tomorrow's lunch

Tide's Out

1. The name of these animals are
   _____.

2. Where do you find them?
   _____

3. What should you carry when searching for quahogs? Why?
   _____
   _____

4. When is the best time to find quahogs?
   _____

   Does it have anything to do with the full moon? *Research.
   _____
   _____

5. Define

   1.) quiver- _____
   2.) quake- _____
   3.) quahog- _____
   4.) fissures- _____

    5.)   siphon- _____

    6.)   invasive- _____

    7.)   ritual- _____

    8.)   discarded- _____

6. What does filter feed system mean?

    _____
    _____
    _____

7. On a sheet of paper draw and colour a quahog shell inside and outside.

8. On a sheet of paper draw the quahog's foot licking the invasive extra salt.

Keep a Treasure Chest of Beachcombing Memories

Find a special box and collect your very own shell treasures.

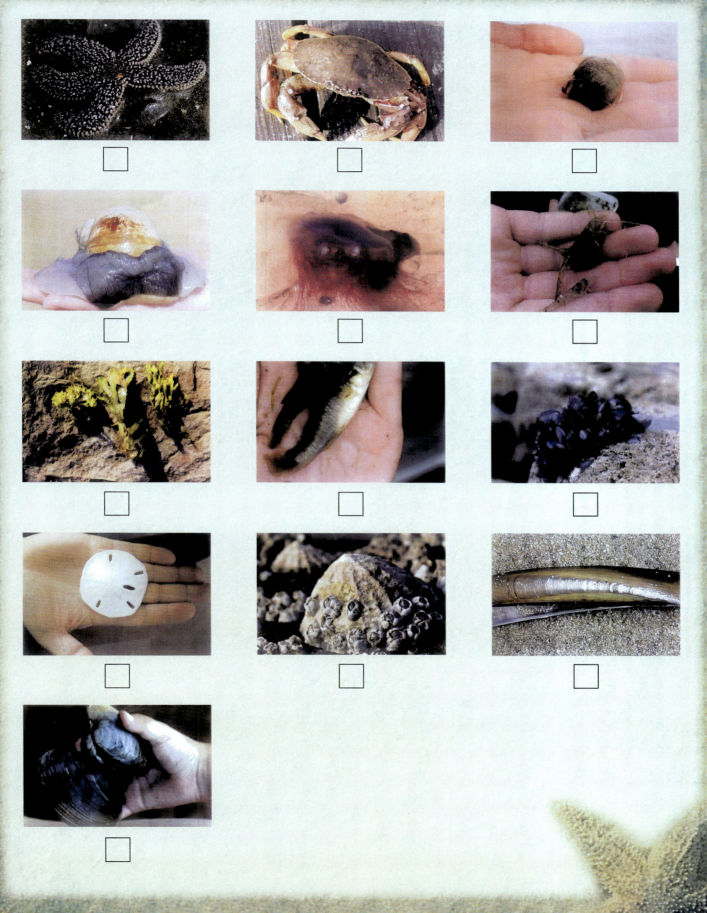

Printed in the United States
by Baker & Taylor Publisher Services